# *The Bible vs. Science?*
# *Or Not!*

By

Joseph McRae Mellichamp

**Thousand Fields Publishing**
www.1000fieldspub.com

Dedicated to scientists throughout the ages
who have sought the truth.

# *The Bible vs. Science?*
# *Or Not!*

# THE BIBLE VS. SCIENCE? OR NOT!

## Introduction

Is the biblical description of the Creation found in Genesis 1:1-2:3 hopelessly at odds with what we know from science about the universe˙and planet Earth? Absolutely not! This may come as a shock to many people because, for reasons which we will discuss, the consistent message communicated by the scientific community, the media, and the education sector is that the account from Scripture is irreconcilable with the story from science. Something I wrote in this connection in 1970 is truer now than it was then.

"Having been educated as an engineer, I had been taught that the Bible is not accurate when it touches on what we have learned from science about our world and the universe. But when I began to check into this, I discovered that the explanation offered by the Bible as to how the universe came into existence fits extremely well with what we know from science, and, furthermore, there are numerous details in the Bible about the Earth and the universe that we are just discovering from science." ["My Search for Success," 1970]

What has happened in the world of science since I wrote these words 40 plus years ago is staggering—the Space Shuttle program, lunar and space exploration, the Space Station, the Hubble Telescope, the computer revolution (from mainframes to PCs), the Internet, and the communication revolution (from analog landlines to digital smart phones) are just a few of the factors which have contributed to our understanding of the universe and our planet. And these factors have clarified and corroborated rather than contradicted the biblical account.

So what is the problem? How has the biblical account come to be held in such low esteem by the scientific community and by the general public and what can be done to change this state of affairs?

Possibly the single most significant dimension of the problem is that, as far as I can tell, there is not a simple, clear statement of how the biblical account meshes with scientific facts. That is, there is no description of how the scientific evidence maps onto the elements of the biblical Creation narrative in a way that is true to science and understandable to the non-scientific inquirer. I intend to provide just such a description in this little book and in the process I will explain how and why the scientific community has not always been forthcoming, how and why the education community has taken the stances it has, and how and why the media portrays things as it does.

In 1986, I presented one of twenty invited papers at a symposium at Yale University on the topic "Artificial Intelligence and the Human Mind." One of the other presenters at that conference was a man named Anthony Flew, one of the most well-known and influential (he says notorious) atheists of our time. Prior to his death in 2010 Flew wrote a book describing his journey from atheism to theism, *There Is A God* [New York: HarperCollins, 2007]. One of the important principles that guided Professor Flew's journey was "the command that Plato in the *Republic* attributes to Socrates: 'We must follow the argument wherever it leads.' (p. 22)"

This will be a good rule for us to follow as we lay out our arguments [evidence]. And we will see that it is precisely because many who engage in this arena do not like where the evidence leads that we have come to the place we now are.

So let's get started. Here is how I am going to proceed. In the biblical account there are seven days of Creation. We are going to divide our discussion into seven chapters, one for each Creation Day. I will take the verses from Genesis 1:1 to Genesis 2:3 that apply to each day and explain them using our current scientific knowledge, actually showing how the scientific evidence dovetails perfectly with the biblical narrative. To help guide our discussion, I have prepared a diagram for each Chapter which plots the various events of that Creation Day on a timeline. I'm sure there will be disagreement over some aspects of my diagrams. I will address most issues of concern in the discussion, so please don't dismiss the material to follow out of hand. Understand my arguments and let them lead where they will.

# DAY 1
## The Creation of the Universe

**The Big Bang**

Genesis 1:1. "In the beginning God created the heavens and the earth."

Most scientists subscribe to some version of what is called the Big Bang Theory. Discoveries in astronomy and physics have demonstrated beyond any reasonable doubt that the universe had a beginning; before that event there was nothing—no matter, no space, and no time. The universe is thought to have begun as an infinitesimally small, infinitely hot, infinitely dense something that immediately began to expand exponentially and continues expanding to the present time.

Whatever the original something was, it included the fundamental building blocks of nature—what physicists call the elementary particles. The laws which govern the elementary particles (motion, gravity, quantum physics, thermodynamics, relativity, etc.) came into existence at this time along with physical space-time dimensions. Cosmogonists, those who study the origin and development of the universe, are reluctant to identify the First Cause behind the Creation event. The Bible in Genesis 1:1 and throughout identifies the First Cause as God and, indeed, it is difficult to imagine *any* reasonable alternative.

Over billions of years, following the laws of nature, the elementary particles combined forming stars, galaxies, planets, moons, cosmic dust (debris), gas, background radiation, black holes, and so forth. The most current estimates suggest that the universe contains 100 to 200 billion galaxies each of which has hundreds of billions of stars. The Milky Way (our own galaxy) has 200 to 400 billion stars. A photograph of just part of our galaxy from the Hubble Telescope is shown on the following page to underscore the absolute enormity of the universe.

1

Hubble Telescope Photo of Segment of Milky Way

The most recent calculation of the age of the universe is 13.73+/-.12 billion years. Scientists are constantly tweaking this number based on analysis of new data, which are coming to light continuously. The current estimate for the age of our Sun is 5 billion years and for our Earth 4.6 billion years. Refer to Figure 1 on page 4 to see the creation of the universe, our Sun and the Earth on an event timeline.

It is at this point that we will have our first vigorous pushback to my explanation from those who subscribe to what is called the Young Earth scenario, believing that the Earth is only about 10,000 years old. I don't wish to argue this point with them; that has been done thoroughly in other venues. I will simply refer them to what John Lennox wrote recently in his book *Seven Days that Divide the World* [Grand Rapids, MI: Zondervan, 2011, p. 62]. Commenting on the geocentric controversy in the time of Galileo (~1610) he wrote, "Only when such a position became mathematically and observationally 'hopeless' should the church have abandoned it." He concludes this is exactly where Young Earth adherents are today, "Young-earth creationism, therefore need not embrace a dogmatic or static biblical [interpretation]. It must be willing to change and admit error."

Before moving on, it will be instructive to make a small digression to comment on the so-called laws of nature mentioned earlier. There is a principle in science called the Anthropic Principle, which essentially says that if any of the laws were changed ever so slightly, life as we know it in the universe would be impossible. [There are actually two cases of the principle—Strong and Weak, but we don't need that level of detail.] You have probably heard of the term "parallel universes." Some scientists propose that there are an infinite number of randomly generated, parallel universes each existing in a slightly different form, and we are just incredibly fortunate to live in the one out of a zillion which supports life. See what's going on here? If our universe had to be finely-tuned ("just so") for life to exist the evidence is pointing toward God as the First Cause and they don't want to go there.

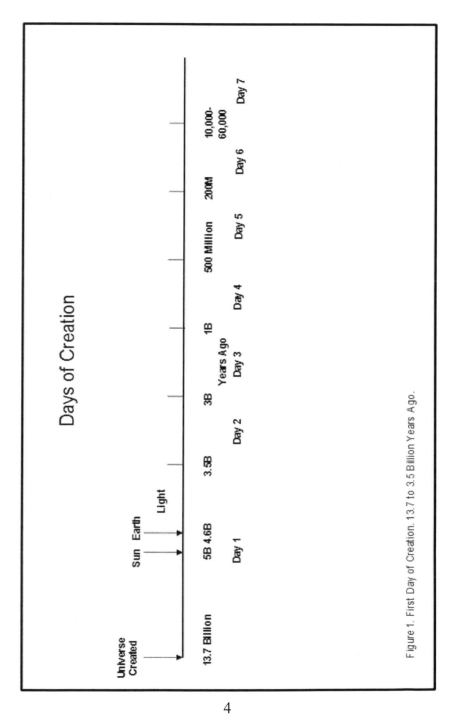

Figure 1. First Day of Creation. 13.7 to 3.5 Billion Years Ago.

By the way, "parallel universes" is not science, it is speculation, because there is no way of proving that there are indeed other universes.

## The Earth

Genesis 1:2. "The earth was formless and void and darkness was over the surface of the deep and the Spirit of God was moving over the surface of the waters."

Many people make a critical error here and are unable to make sense of much of what follows because they don't understand that the frame of reference shifts between verse 1 and verse 2 of Genesis 1. Genesis 1:1 is about the Creation of the universe—it is describing a cosmic event. Genesis 1:2 is focused on the Earth, thus, some 9 billion years had elapsed from God's original Creation act. Stars, planets, and galaxies have formed, and ceased to exist even as they are doing to the present time. What follows in the Genesis narrative is a description of God's creative activity on planet Earth.

We have learned much about how planets form since the Hubble Telescope was deployed in 1990. Astronomers now believe that stars and their planets form out of a collapsing cloud of dust and gas within a larger cloud called a nebula. Some of the material from the nebula is compressed together becoming increasingly hotter until it becomes the nucleus of a star; other material from the nebula begins to accumulate into smaller clumps which eventually become planets orbiting the star. All of this is taking place within the nebula, thus the emerging planets would be surrounded by thick layers of gas, dust, and debris.

When the writer of Genesis [most believe it was Moses writing around 1400-1450 BC] describes the Earth as "formless and void and darkness was moving over the surface of the deep," he was likely describing the initial conditions on the Earth, prior to when it matured as a planet. Of course, other stars existed at this point in

5

time, and our own sun would also have been emitting light. The point is that the Earth was so shrouded by the gas, dust, and debris, that an observer on the Earth's surface at that time would have been unable to detect light from the sun or other stars.

## Light

Genesis 1:3-5. "Then God said, 'Let there be light'; and there was light. God saw that the light was good; and God separated the light from the darkness. God called the light day, and the darkness He called night. And there was evening and there was morning, one day."

Something happened between Genesis 1:2 and Genesis 1:3 so that light would have been visible on the surface of the Earth. And what happened scientists believe was the formation of our Moon. Here is how they think it happened, according to a theory called the "Giant Impactor Hypothesis." Approximately 4.25 billion years ago, a smaller planet about the size of Mars, which had formed in a process similar to Earth's and in an orbit intersecting with Earth's, collided with the Earth in a not quite head-on collision. In the collision, the core of the smaller planet was almost totally absorbed into the Earth's core, and huge amounts of the mantle of both earth and the smaller planet were blasted into space along with much of the gas/debris layer covering the Earth. This material then in time coalesced to form our Moon.

Lunar samples brought back from the Apollo missions suggest the Moon is 4.25 billion years old. The age of the Earth is thought to be about 4.6 (actually 4.59) billion years. These dates along with comparisons of the compositions of the Earth and Moon provide strong evidence for the Giant Impactor Theory. After the collision, it is thought that the Earth's atmosphere would have been translucent (like the cloud cover on a very overcast day), but thin enough to permit the passage of light, thus making it possible to differentiate day and night from the surface of the Earth for the first time in its lifetime just as the Genesis account describes.

The biblical account for this part of the Creation ends with the phrase, "there was evening and there was morning, one day," a pattern that the writer of Genesis uses to set apart the Creation Days. Thus, according to our timeline the First Day of Creation was approximately 10.2 (13.7-3.5) billion years in duration.

## DAY 2
### Water and the Expanse

Gen 1:6-8. "Then God said, 'Let there be an expanse in the midst of the waters, and let it separate the waters from the waters.' God made the expanse, and separated the waters which were below the expanse from the waters which were above the expanse; and it was so. God called the expanse heaven. And there was evening and there was morning, a second day."

The biblical narrative informs us that on Day 2, God began putting the Earth's water or hydrologic cycle in place. The water cycle is what we call the process by which water is constantly being cycled between the Earth's surface and the Earth's atmosphere—the layer of gases surrounding the Earth and held in place by its gravity. We are told that God caused a separation between what we call the oceans and other surface water ("waters under the expanse") and the troposphere ("water above the expanse")—the expanse including what we call the atmosphere. Technically, the troposphere is the lowest layer of the atmosphere and contains 80% of the atmosphere's mass and 99% of its water vapor and aerosols (particles).

Of course water is an essential element for life on Earth. It is necessary in a multitude of ways to sustain plant and animal life, and, in turn, human life. Ever notice how the first thing space explorers look for on other planets is the presence of water? No water, no life. Earth's water cycle provides for a stable supply of water as it is constantly being cycled between the Earth's surface and the atmosphere in a complex process powered by energy from the Sun. Water from the oceans and the land surfaces evaporates, the vapors rise into the atmosphere, join with dust particles, condense into clouds and then water falls back to the surface in the form of rain, ice, or snow.

In his book *The Genesis Question*, Hugh Ross shares insight on the fine-tuning necessary to balance this amazing system

[Colorado Springs, CO: NavPress, 1998, pps. 35,36]. Recent research has uncovered what could represent a deadly imbalance in the water cycle, and at the same time, the remarkable phenomenon that compensates for it. Complex life processes result in an ongoing conversion (loss) of water from the water supply. And a small amount of water escapes Earth through gravitational dissipation. An on-going influx of water-rich extraterrestrial material—comets of all sizes (regular, small, mini and micro) collide with the atmosphere contributing to the water supply just enough to offset losses. This phenomenon, unknown until the 1980s and unproven until the late 1990s, is just one of numerous "just so" occurrences which allow complex life on our planet.

The writer of Genesis tells us that God called the expanse heaven. In this context, heaven is everything in the universe except the Earth; thus heaven includes the Earth's atmosphere, the rest of our solar system—the Sun, moons and planets—plus everything else. I suggest in Figure 2 on page 10 that this activity took place in the interval between 3.5 and 3 billion years ago. Just to underscore that these dates are approximations, here is a funny story that actually happened. I was sharing these concepts with a small group several years ago and one of the group members asked in a very agitated state, "Well, I just want to know when the dinosaurs came on the scene [a topic we will answer for certain when we get to Days 5 and 6]!" I replied, "Ellen, God created dinosaurs on Thursday morning at 11:00 am." Remarkably, she was satisfied, even though everyone knew I was exaggerating to make a point. And the point is that we can't be precise in dealing with such vast time scales; we must deal with approximations. The intervals I've shown in the diagrams are approximate; I have labeled them according to my best understanding of when the scientific data appear. My friend's question did spur me to do some further digging and I uncovered the material on dinosaurs which will be presented in those later chapters.

9

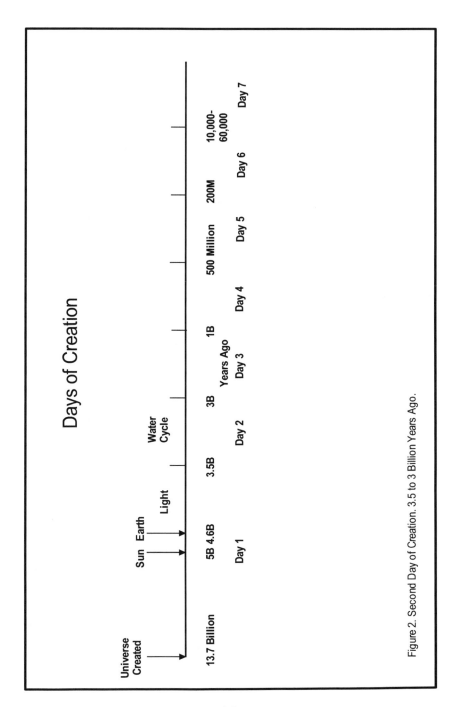

Days of Creation

Figure 2. Second Day of Creation. 3.5 to 3 Billion Years Ago.

# Day 3
## Continents and Plants

Genesis 1:9-13. "Then God said, 'Let the waters below the heavens be gathered into one place, and let the dry land appear'; and it was so. God called the dry land earth, and the gathering of the waters He called seas; and God saw that it was good. Then God said, 'Let the earth sprout vegetation, plants yielding seed, and fruit trees on the earth bearing fruit after their kind with seed in them'; and it was so. The earth brought forth vegetation, plants yielding seed after their kind, and trees bearing fruit with seed in them, after their kind; and God saw that it was good. There was evening and there was morning, a third day."

## Continents

Next the biblical narrative tells us that the waters "below the heavens," that is on the Earth, were gathered and dry land appeared. Obviously land is necessary for all life except aquatic life. Hugh Ross asks the question, "What does geology tell us? ...for the first four billion years of Earth's history, the landmass grew from 0 percent of the planetary surface to 29 percent [*The Genesis Question*, pps. 38]." Scientists are uncertain how or when this occurred; most geologists believe some form of the Continental Drift Theory which states that at one time all the continents were joined in a super-continent, which they call Pangaea. Over a vast period of time, the continents drifted apart to their current locations.

Since the appearance of the continents is described in the biblical narrative in conjunction with the first appearance of life on Earth, I have shown it in Figure 3 on page 12 in the interval from 3 to 1 billion years ago, but acknowledge that it could well have overlapped this interval. An animation of what the landmass might have looked like over time is shown on page 13; the actual time of the animation as shown is considerably more recent than what I have shown on my diagram.

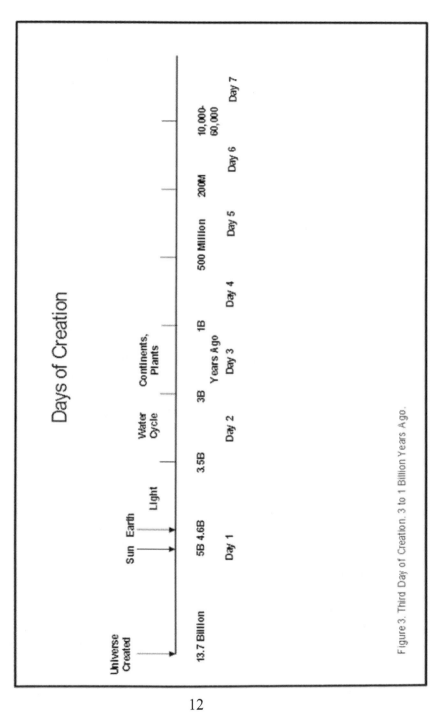

Figure 3. Third Day of Creation. 3 to 1 Billion Years Ago.

# Animation of Continental Drift

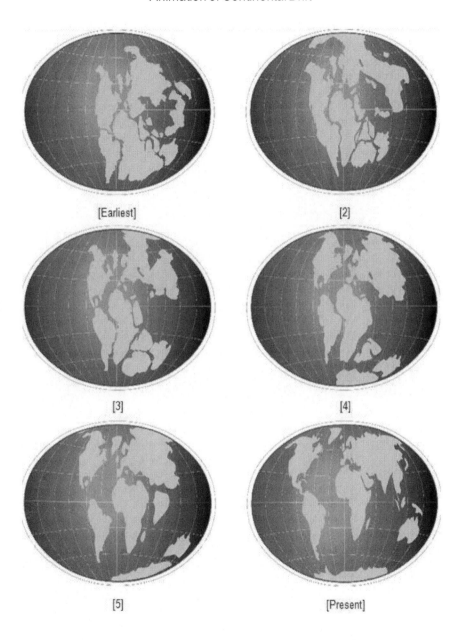

[Earliest]

[2]

[3]

[4]

[5]

[Present]

## Plants

When the landmass was ready to support life, the biblical narrative tells us that, "The earth brought forth vegetation, plants yielding seed after their kind, and trees bearing fruit with seed in them, after their kind." What do scientists believe about the origin of life on Earth? A couple of quotes from John Lennox's book *God's Undertaker: Has Science Buried God?* [Oxford, England: Lion Hudson plc, 2007] are in order here:

- "Anyone who tells you that he or she knows how life started on the earth some 3.45 billion years ago is a fool or a knave. Nobody knows." [Lennox quoting Stuart Kaufmann, p.116].

- "At present all discussions on principal theories and experiments [on the origin of life] either end in stalemate or a confession of ignorance." [Lennox quoting Klaus Dose, p.126].

The reigning theory at the moment is a modernized version of Panspermia, which holds that life on Earth was seeded from space, and that life's evolution to higher forms depends on genetic programs that come from space. If this were not such a dangerous idea, it would be laughable. Scientists know that the possibility of life spontaneously generating on Earth is nonexistent. Unfazed they concoct this theory that it originated in outer space—never recognizing that this simply removes their problem to another location without adding one iota of understanding to the puzzle.

It reminds me of a cartoon, *Non Sequitur*, by Wiley, that appeared in our Sunday newspaper on August 18, 2013 which I have reproduced on page 16. The first panel, captioned "Capt. Eddie's Apocryphal Science," shows two space aliens in their spacecraft, presumably flying over planet Earth. The next panel shows one of the two depressing a red button titled "Gloop Pump." This is followed by an exterior shot of the spacecraft spewing bilious liquid as it passes over the planet; the liquid falls into a primordial lake in the fourth panel. Next we see microscopic

14

organisms emerge followed by what appears to be a tadpole, and then a fish. Then we are treated to the ubiquitous fish becomes four-legged critter becomes ape becomes man depiction. In the final panel Captain Eddie says, "It's called the Unintended Consequences Theory." Captain Eddie, I give you an "A" in biology because you nailed the prevailing theory of the origin of life!

I'll leave it to you, Reader. Which is more plausible: the Science/Non Sequitur scenario, life came from outer space; or the biblical statement, "God created it?" I don't know about you, but it is an easy conclusion for me to draw.

Regardless of how life came to be on planet Earth, the oldest fossils are 3.5 and 3.4 billion year old single-cell, blue-green algae found in Western Australia and South Africa. The biblical narrative states, "'Let the earth sprout vegetation, plants yielding seed, and fruit trees on the earth bearing fruit after their kind with seed in them'; and it was so. The earth brought forth vegetation, plants yielding seed after their kind, and trees bearing fruit with seed in them, after their kind." Simple single cell life existed in the 3.5 billion year timeframe; complex vegetation including plants and trees are on the planet in the 3 to 1 billion year timeframe as shown on the diagram on page 12.

## Non Sequitur by Wiley
8-18-2013

# DAY 4
## Cloud Layer

Genesis 1:14-19. "Then God said, 'Let there be lights in the expanse of the heavens to separate the day from the night, and let them be for signs and for seasons and for days and years; and let them be for lights in the expanse of the heavens to give light on the earth'; and it was so. God made the two great lights, the greater light to govern the day, and the lesser light to govern the night; He made the stars also. God placed them in the expanse of the heavens to give light on the earth, and to govern the day and the night, and to separate the light from the darkness; and God saw that it was good. There was evening and there was morning, a fourth day."

As we saw when we were discussing the events of Day 1, the atmosphere of the early Earth was opaque, that is so cluttered with gases, dust, and debris that it was impossible for light from the sun or stars to penetrate. Thus, had an observer been on the surface of the earth [of course no such observer was present at the time for life of any kind was a billion years in the future], everything would have been clothed in darkness, just as Genesis 1:2 declares. We saw that the creation of the Moon involved a violent collision of the Earth with a smaller planet and the outcome of this was that the atmosphere gradually changed from opaque to translucent. There was still a heavy cover surrounding the Earth, but light was able to penetrate.

Over time, the atmosphere gradually cleared owing to several factors which Hugh Ross identifies as changes in temperature, pressure, volcanic activity, rotation speed, wind velocities, and oxygen diffusion [*The Genesis Question*, pps. 42-44]. Thus at some point in time it would have been possible to observe the sun, moon and stars from the surface of the Earth. Genesis 1:14-19 describes that this happened. Unfortunately, many people have misunderstood these verses because of the verb construction from the Hebrew language in which the passage was originally recorded. Here is a simple explanation of the problem.

17

Hebrew verbs did not have tenses that communicate information about the time sequences of other events as many languages do. Thus any information relative to the timing of other events when one is dealing with the Hebrew past tense must be inferred from the context and the grammar of the text. Here we have in Genesis 1:16 the statement "God made the two great lights ... He made the stars also." In English, this seems to be a simple statement of past action. Some read this and conclude the narrative is stating that God made the Sun, Moon and stars on Day 4 after He made the Earth.

But the context and grammatical construction of the narrative don't support such an assertion. In English, we would use the pluperfect tense for this statement, thus it would read, "God had made..." And we would understand the narrative to be saying that at some point in time after God had made the Sun, the Earth, the Moon, and the stars, the atmosphere of the Earth cleared enough so that these bodies were visible from the surface of the Earth, and that at some future point after the advent of humans, they would make use of these bodies for light and for demarking day and night and the different seasons of the year.

We have shown this activity in Figure 4 on page 19 in the interval from 1 billion to 500 million years ago and labeled "The Cloud Layer." It should be underscored that this clearing of the atmosphere was a process which would have occurred gradually over time. On an overcast day, sometimes clouds move out fairly quickly. When the entire atmosphere of the Earth was loaded with cosmic junk, it would take many millions of years for the transition to occur.

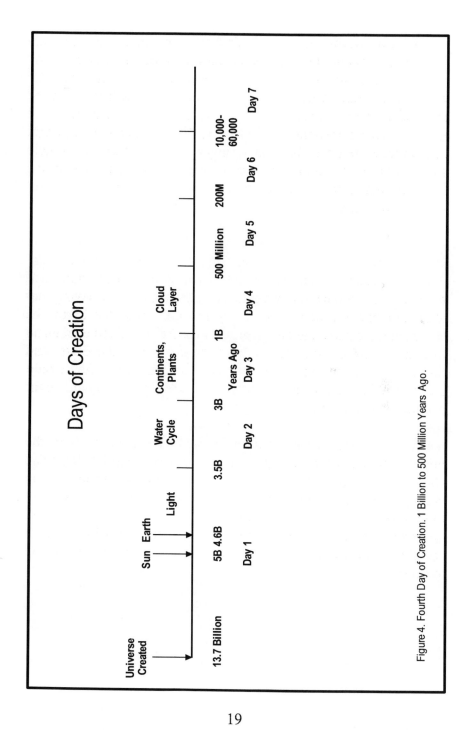

Figure 4. Fourth Day of Creation. 1 Billion to 500 Million Years Ago.

## DAY 5
### Sea Animals, Sea Mammals, Birds

Genesis 1:20-23. "Then God said, 'Let the waters teem with swarms of living creatures, and let birds fly above the earth in the open expanse of the heavens.' God created the great sea monsters and every living creature that moves, with which the waters swarmed after their kind, and every winged bird after its kind; and God saw that it was good. God blessed them, saying, 'Be fruitful and multiply, and fill the waters in the seas, and let birds multiply on the earth.' There was evening and there was morning, a fifth day."

At Day 5, the biblical narrative informs us of the creation of sea animals and birds. And we are told that they were created after their kind, that is, species by species. What does science tell us about the creation of living organisms? Many scientists today subscribe to some version of the Theory of Evolution—the idea that all species of life have evolved over time from a common ancestor. Most versions of the theory state that an unguided (or purposeless) process of random mutation and natural selection is responsible for the various life forms evident on Earth today or in the fossil record. The most common way of explaining what the process entails is to present a depiction of what is called "Darwin's Tree." An example is shown in the upper part of the diagram on page 23. Does this really explain what happened?

Consider another quote from John Lennox's book **God's Undertaker**: "We paleontologists have said that the history of life supports [the story of gradual adaptive change] knowing all the while that it does not." [Niles Eldredge, quoted in **God's Undertaker**, p.111] Niles Eldridge is an American biologist and paleontologist who along with Stephen Jay Gould proposed the Theory of Punctuated Equilibrium in 1972. Eldredge and Gould realized that the earlier, simplistic notion of gradual change over time wasn't always reflected in the fossil record. They claimed that stasis [a period or state of inactivity] for eons is the norm, "punctuated" by brief periods of rapid evolutionary change.

Is this true science, or is it a clever way of getting around the embarrassing fact of the utter lack of examples of transitional fossils [any fossilized remains of a life form that exhibits traits common to both an ancestral group and its derived descendant group] in the fossil record? John Lennox is a great deal more critical when he asks, "But why? What conceivable reason could there be for members of an academic community to suppress what they know to be the truth—unless it were something which supported a worldview, which they had already decided was unacceptable." [*God's Undertaker*, p.111]

So what is the truth? Did all life forms which have existed on Earth come from a common ancestor and evolve over eons of time through an unguided process of random mutation and natural selection—punctuated equilibrium or not? Or is there another scenario which explains the scientific data more exactly?

Several years ago, Stephen Meyer published a peer reviewed article, "The Origin of Biological Information and the Higher Taxonomic Orders," in a scientific journal. In the article Meyer states that about 500 million years ago, in a narrow 5 to 10 million year window of geologic time, at least 19 and perhaps as many as 35 of the total 40 phyla make their first appearance. This is called the Cambrian Explosion. In almost all cases the Cambrian animals have no clear morphological antecedents (animals with similar body plans) in earlier Precambrian fossils. [Meyer, Stephen C., "The Origin of Biological Information and the Higher Taxonomic Orders," *Proceedings of the Biological Society of Washington*, September 2004.] What this would look like is shown on the lower part of the diagram on page 23.

Now these two possibilities, common descent and Cambrian Explosion, could not be further apart. Many, if not most, biologists will grudgingly admit to the factual basis of the Cambrian

21

Explosion; they are less forthcoming as to the implications of the Cambrian fossils for the idea of common descent. You be the judge. Look again at the diagrams on page 23. The one on the lower portion of the page is a fairly true representation of the geological evidence; the one on upper portion of the page can only be called what it is, a figment. Here are two additional insights that should help your decision:

The editor of *Proceedings of the Biological Society of Washington,* Richard Sternberg, lost his job over his role in publishing the Meyer article, even though he followed all of the publication's guidelines in requiring the piece to work its way through every step of the review process. What was going on here? Reminds one of Lennox's earlier comment, "What conceivable reason could there be for members of an academic community to suppress what they know to be the truth—unless it were something which supported a worldview, which they had already decided was unacceptable."

Shortly after the Meyer article was published, a faculty group at the Georgia Institute of Technology with which I was involved, invited a biology professor from another, well-respected university to do a presentation in which he would comment on Meyer's article. As the professor started his talk, he berated the group for not buying biology's "party line," that is, common descent. Then, he showed his first PowerPoint slide—a graphic of Darwin's tree! I nearly lost it. I immediately raised my hand and interrupted him. "Professor X," I said, "You know very well that if what Dr. Meyer has written in his article is true, Darwin's Tree is a figment of your imagination. But just to be fair, would you please give us one example of a transitional animal from the fossil record. I don't need fifty examples, I don't need ten. I just want one that is generally accepted and not disputed." He mumbled a few words under his breath and tried to change the subject. I wouldn't let him. He finally said that he could not.

# THE DAYS OF CREATION
## Day Five

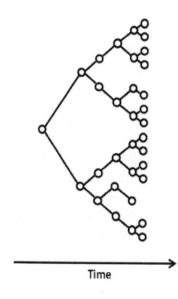

### Darwin's Tree
How the fossil record would appear based upon Darwin's assumption of a common ancestor. Many biology text-books show a diagram like this.

Time

### Cambrian Explosion
How the fossil record actually appears showing species appearing (and disappearing) and remain-ing relatively stable thru time.

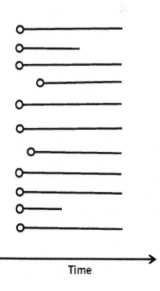

Time

Something is terribly wrong here. It is not the biblical explanation of how life came to be on Earth; it is the dishonest way the scientific community has chosen to interpret and teach the scientific evidence that relates to the origin of life.

Accepting the generally recognized date of 500 million years ago as the likely beginning of the Cambrian Explosion, I have shown the creation of sea animals and birds in the interval from 500 to 200 million years ago in Figure 5 on page 26. The earliest bird fossils date to 160 million years ago; insects, which along with seeds, are the main diet of birds date to 300 million years ago.

Before wrapping up Creation Day 5 and moving to Day 6, I want to mention three animals of great interest to many people: dinosaurs, and "Behemoth" and "Leviathan," the latter two of which are mentioned in chapters 40 and 41 of the Book of Job in the Bible.

Although Scripture doesn't mention them, dinosaurs existed in the Mesozic (Reptile) Era 225-65 million years ago. For fans of the movie *Jurassic Park*, the Jurassic Period is 195-136 million years ago. Accordingly, I show dinosaurs on Figure 5 just at the end of Day 5 and overlapping into Day 6. It is believed that an asteroid collision with the Earth 61 million years ago resulted in the extinction of from 1/2 to 2/3 of all species on the planet, likely including dinosaurs. Just to finish the story, the Creation of Man we will see is at the very end of Day 6 in the interval 60,000 to 10,000 years ago; thus, sadly, the really scary scenes in *Jurassic Park* of humans being chased madly through the park by dinosaurs looking for a quick bite on the run never happened. Dinosaurs were long gone by the time humans arrived on the scene. Sorry to spoil the fun, Steven Spielberg.

Finally, "Behemoth" and "Leviathan" of Job 40,41 probably refer, not to dinosaurs, but to hippopotami and crocodiles respectively. The former are land mammals and are among those created during Day Six. Ever wonder what would happen if

Behemoth and Leviathan fought it out? Check out the photo on page 27.

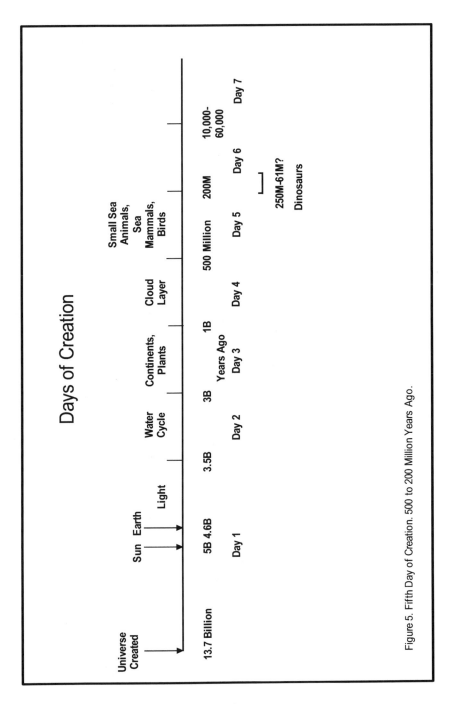

Figure 5. Fifth Day of Creation. 500 to 200 Million Years Ago.

27

## DAY 6
## Land Mammals, Man

Genesis 1:24-25. "Then God said, 'Let the earth bring forth living creatures after their kind: cattle and creeping things and beasts of the earth after their kind'; and it was so. God made the beasts of the earth after their kind, and the cattle after their kind, and everything that creeps on the ground after its kind; and God saw that it was good."

Here we are told that on Day Six God created three different kinds of land mammals: cattle or livestock (the agricultural animals that could be domesticated), creeping things (these would include what we would call smaller wild animals), and the beasts of the Earth (these would include most of the larger wild animals). The list refers to general categories; specific species of animals are not given. We show these creatures in Figure 6 on page 31 between roughly 200 million years ago to about 10,000 years ago. Fossil remains of domesticated animals used in dairy farming have been found dating to 10,000 years ago and some dating much earlier. Remains of smaller animals dating to 200,000 years ago have been reported, again some much earlier. And finally, fossil remains of wild animals (lions and tigers) dating to 1 to 2 million years ago appear in the fossil record.

Genesis 1:26-27, 31. "Then God said, 'Let Us make man in Our image, according to Our likeness; and let them rule over the fish of the sea and over the birds of the sky and over the cattle and over all the earth, and over every creeping thing that creeps on the earth.' God created man in His own image, in the image of God He created him; male and female He created them. God saw all that He had made, and behold, it was very good. And there was evening and there was morning, the sixth day."

In Genesis 1:26-27 and 31, we have our first glimpse of the Triune God—Father, Son and Holy Spirit. "Our" is plural while "He" and "His" are singular. Hugh Ross [*The Genesis Question*,

p. 54] observes that from Scripture and from study of human culture, we understand that to be created in the image of God implies having a spirit—this is what separates humans from all other creatures. This includes the following:

- Awareness of the moral law.
- Concerns about the afterlife.
- A desire to worship a higher being.
- Consciousness of self.
- An understanding of truth.

From Biblical genealogies, which we know are incomplete, it is estimated that man was created from 10,000-60,000 years ago. Thus, this date is shown on the diagram on page 31 for the culminating Creation event on Day 6—the Creation of Man.

From archaeology, remains of human beings (i.e., spiritual beings or those who left evidence of the attributes enumerated in the second bullet above) date to 8,000-40,000 years ago. And some of the earliest examples we have are:

- Cro-Magnon Man (30,000BC) found in France was among the first fossils recognized as (Homo-sapiens) modern man.
- The cave paintings of Lascaux, France, which reflect the attributes of humans are dated from 10,000-25,000 years ago.

Now what about the class of beings that scientists designate as Hominids? Are they humans or not? Fazale Rana and Hugh Ross have done a masterful job of noting the differences between the hominids and modern man. [*Who Was Adam?* Colorado Springs, CO: NavPress, 2005]. Based on a number of different factors including DNA they convincingly argue that the Hominids are distinctly different from modern man.

From archaeology, Bipedal Primates or Hominids (*Australopithecus*, *Homo erectus*, etc.) date to 1-4 million years ago. Evidence suggests that the Bipedal Primates, with the possible exception of Neanderthal Man, went extinct before the advent of humans. Hominids are shown on our Creation timeline on page 31.

The prehistoric types, i.e., Neanderthal Man (230,000-30,000BC?), Java Man (50,000-10,000BC), and Peking Man (50,000BC) are probably of the *Homo erectus* category. Piltdown Man was a clever fake consisting of part of a human skull and the jawbone of an ape.

Finally, speaking of the *Homo erectus* fossils, William Kimbel, director of the Human Origins Institute at Arizona State University says, "There are only a handful of specimens. You could put them all into a small shoe box and still have room for a good pair of shoes." [*National Geographic*, August 2011, p. 132] An illustration of "artistic liberties" taken in reconstructing Neanderthal Man from his skull fossils is shown on page 32. Note the progression from the first reconstruction on the left to the one on the top.

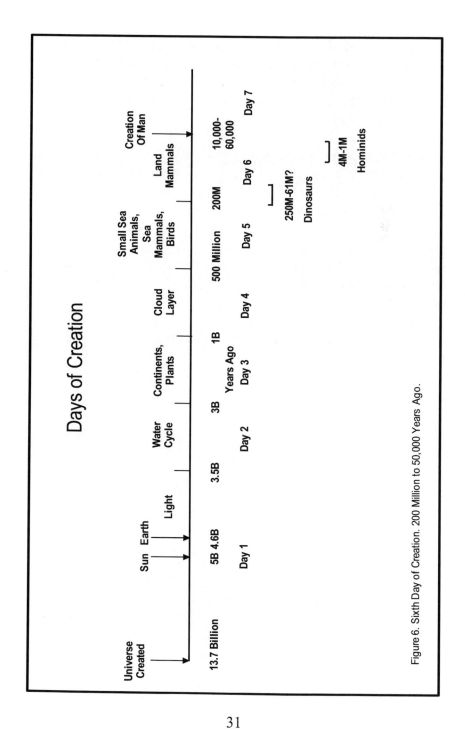

Figure 6. Sixth Day of Creation. 200 Million to 50,000 Years Ago.

First reconstruction of Neanderthal man

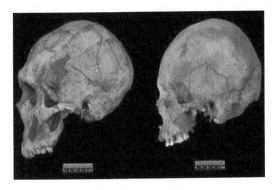

# Neathandertal Man

## DAY 7
## God Rested

Genesis 2:1-3. "Thus the heavens and the earth were completed, and all their hosts. By the seventh day God completed His work which He had done, and He rested on the seventh day from all His work which He had done. Then God blessed the seventh day and sanctified it, because in it He rested from all His work which God had created and made."

These verses in the Creation account need no explanation—they are clear and unequivocal. The take away here is that if God rested after His Creative work, one day of rest in seven is probably a good rule of thumb for Man, God's Creation. Thus, these verses are the basis for the Fourth Commandment. A number of years ago I read A. B. Bruce's classic, *The Training of the Twelve* [Alexander Balmain Bruce, *The Training of the Twelve*, Grand Rapids, MI: Kregel Publications, 1988]. Bruce, a Scottish clergyman writing in 1871 said of the Sabbath, "It was not a day taken from man by God with a demanding spirit, but a day given by God in mercy to man—God's holiday to His subjects. ... The best way to observe the Sabbath is that which is most conducive to man's physical and spiritual well-being—in other words, that which will be best for his body and soul. In the light of this principle, you will keep the holy day in a spirit of intelligent joy and thankfulness to God the Creator for His gracious consideration toward His creatures." Amen!

Day 7 is shown in our timeline on the following page.

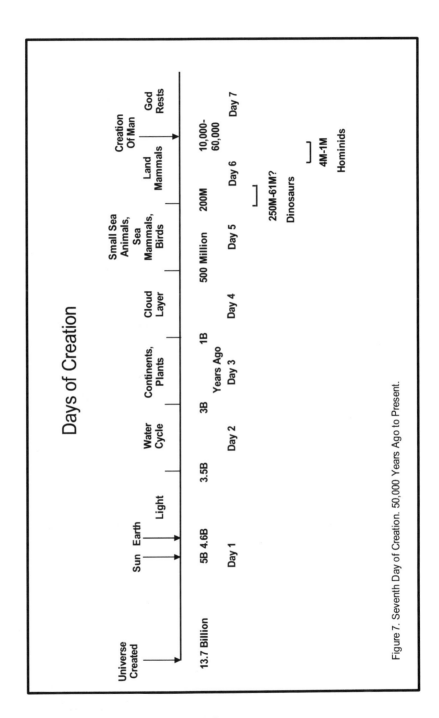

Figure 7. Seventh Day of Creation. 50,000 Years Ago to Present.

34

# THE BIBLE VS. SCIENCE? OR NOT!

## Epilog

Well, there you have it. It is very clear to me and has been for many years that the biblical account of the Creation is absolutely supported by what we know from science about these events. In fact, I would even say that the biblical account is perfectly supported by the scientific evidence. I hope as you have read my attempt to harmonize the scientific evidence with the verses in Genesis 1:1-2:3, you agree with my assessment. Rather than being hopelessly at odds with the scientific facts, we should marvel at the magnificent insight into His Creative work the Creator has given us in this incredible book (Genesis) written 3,400+ years ago!

In the Introduction, I wrote that in addition to showing how the scientific evidence fits the biblical narrative of the Creation, I wanted to show "how and why the scientific community has not always been forthcoming, how and why the education community has taken the stances it has, and how and why the media portrays things as it does." It should have become apparent now that the scientific community has misrepresented the evidence to support its particular worldview, and that the reason it has done this is because the only other alternative (the biblical worldview) is unacceptable.

What about the education community and the media? Why have they so faithfully followed the "party line" when it comes to the Creation? The answer should be obvious. Educators and media workers (and scientists too for that matter) are, for the most part, educated in the universities of the world, and the reigning worldview in the university is that of scientific materialism—the belief that physical reality, as made available to the natural sciences, is all that truly exists. Having spent over fifty years of my life in the university as a student, a professor, and a writer and speaker, I know how difficult it is to go against the tide and to embrace ideas that are counter to the mainstream. But we must follow the evidence where it leads. ***Truth is the goal!***

Earlier, I wrote that I had presented a paper at Yale University in 1986 at a conference, "Artificial Intelligence and the Human Mind." As I wrap up this work, I am reminded of a quote I discovered as I did the research for that paper by a computer science professor from the Massachusetts Institute of Technology. It is appropriate here. "It is not those of us who seek to understand the world from a number of different perspectives, including the scientific one, who prefer ignorance to knowledge. It is those who, blinded by their faith that science can yield 'full' explanations, prefer to remain ignorant of whatever knowledge other ways of knowing the world have to offer." [Joseph Weizenbaum. "The Last Dream." ***Across the Board***, Vol. 14, No. 7 (July 1977), pp. 34-46.]

Professor Weizenbaum has it exactly right. Science is a wonderful thing! It has contributed to our lives in innumerable ways. But even science has its limitations. It cannot answer the significant questions of life: Why is there something rather than nothing? Why are we here? Where are we going? These are metaphysical questions—they are beyond science. We have seen that the Bible and science align perfectly when they address the questions related to how and when the universe and the things in it came to be. Thus, we must follow Professor Weizenbaum's advice; we must turn to another perspective [source] for answers to the questions science can't answer. The Bible provides meaningful answers to the significant questions of life—the ones we must answer if we are to live and finish well.

Made in the USA
Charleston, SC
08 January 2014